The Queen of the Night:

The Night-blooming Cereus

by Elaine A. Powers

Illustrated by Nicholas Thorpe

Queen of the Night: the Night-blooming Cereus

by Elaine A. Powers
Copyright 2019 by Elaine A. Powers
All rights reserved.

ISBN-13: 978-0-9991669-6-3

Published by Lyric Power Publishing LLC, Tucson AZ

The *Reina de la Noche* or Queen of the Night
Finds southwest desert habitats just right.
Cereus like growing with ironwood and creosote,
That shade, support, conceal and promote.

2

They drink little water and climb really slow;
In full sun or partial shade, Cereus can grow.
But when the right conditions inspire
A stalk can grow ten inches or higher.

Cereus are hardy, desert living makes them so,
But they can be damaged if temperatures go too low.
When these cacti find a location to thrive,
One hundred years later, they may still be alive.

The stem is one half to a whole inch wide,
With six to nine edges on each side.
You'll notice, they have no leaves on them;
All photosynthesis takes place inside the stem.

In winter, the plant dies back to the ground;
Underneath is where its large tuber is found.
It can be forty pounds and tastes like potato,
People can eat them and that's good to know.

Boiled roots were once used for medicine
To help breathing, headaches and digestion.
Other animals eat the tubers as well;
Javelinas and pack rats think they taste swell.

Under the heat of the warming desert sun,
The swelling of little buds has now begun.
This moment comes in late April or early May,
When the temperatures increase during the day.

But the flowering progress is not steady or sure,
The buds' growth starts and stops as they mature.
The buds form on areoles each year;
Anywhere from the top to midway down, they may appear.

If the fuzzy bud is gray, it becomes a new arm,
If pink, it will become a flower, the cactus' charm.
A fuzzy pink bud swells and grows steadily,
Then stalls – less than an inch is all it will be.

For one day to four weeks, the bud stops its growing
'Til the others catch up – then they really get going.
Once they break stall, they're committed to bloom.
People wait with excitement - flowers will be here soon!

When will the flower bloom on the simple stick?
The size of the bud will help you predict.
At 60 millimeters long, it's still a week,
But at 80 millimeters, in four to five days it will peak.

At 120 millimeters, blooming is a few days away;
At 170-230 millimeters, there'll be no delay.
Of course, some Cereus buds don't care about the date,
They might bloom two days early to a whole week late.

How do they know when their neighbors will show?
Many things happen to tell them so.
Cues in the environment bring out their flowers,
Like pressure, humidity and temp changes after showers.

8

They'll flower in July or as early as late June,
Before the start of the season called desert monsoon.
When it's time to open, the petals loosen their hold
Which means the mass blooming is about to unfold.

The bud unfurls to a flower the size of a hand,
And some span ten inches wide, isn't that grand?
The many white petals stretch out and open wide
To reveal the contrasting yellow stamens inside.

9

They open after sundown but soon they are gone;
The end of the spectacle comes with the dawn.
Yes, these beauties will bloom only at night.
Their white flowers glow in the silver moonlight.

Some stems only have one or two flowers,
While some have forty, creating quite the bower.
Not all the flowers look the same as you may think,
Sepals can be red, emerald, peridot, or pink.

Not only that! The stems are often dark green,
But purplish-pink stems may also be seen.

The fragrance of vanilla perfumes the air;
The intensity is due to the flowers fair.
Their scent signals the pollinators,
Telling them to come sooner, rather than later.

11

Who pollinates the flower, you may ask?
Bats and sphinx moths handle the task.
Sphinx moths look a lot like hummingbirds;
When they fly, a loud humming sound is often heard.

From a half mile away, the plant's pheromone
Attracts the pollinating moths as they roam.
They hover above with their three-inch wings,
Which flap so rapidly they seem to sing.
With their long tube-like tongues uncoiled, they feed,
Collecting pollen, as they sip the nectar they need.

Nectar-feeding bats have different styles;
Some use a long tongue, but others land to drink a while.
But not all pollinators come at night,
Bees visit the blooms at the first morning light.

In a month or two, the ripe Cereus fruit turns deep red;
Birds eat them and that helps the seeds spread.

Despite the birds' help, the numbers of Cereus are decreasing.
It's possible that nights of the Queen could be ceasing.
Their survival is threatened by collectors and farms.
Pesticides kill the moths and agriculture causes other harms.
We know that plants can be grown from cuttings or seed;
Don't dig them up! There is simply no need.

15

The flowering of the Night-blooming Cereus is an incredible sight,
But remember it only happens on one special night!

17

Glossary

Areole: A spot on the cactus from where spines and flowers grow. Areoles are modified branches and considered an identifying feature of cacti.

Cactus: Cacti are succulent, perennial plants. Known for their thick chlorophyll-containing stems, cacti are distinguished from other succulent plants by the presence of areoles. In most cacti, leaves are absent, reduced, or modified as spines, minimizing the amount of surface area to reduce water loss. Their shallow root systems are designed to absorb any available moisture.

Millimeter: A Metric unit of length. Twenty-five millimeters equal an inch.

Photosynthesis: The process plants use to convert sunlight to carbon dioxide and water into food. Oxygen is generated as a byproduct. The green pigment chlorophyll is usually involved.

Sepals: These modified leaves protect the flower when it is in bud. They also provide support for the petals when the flower is in bloom.

Stall: A bud stops growing and is considered being in a "stalled status." The bud "breaks stall" when it starts growing again.

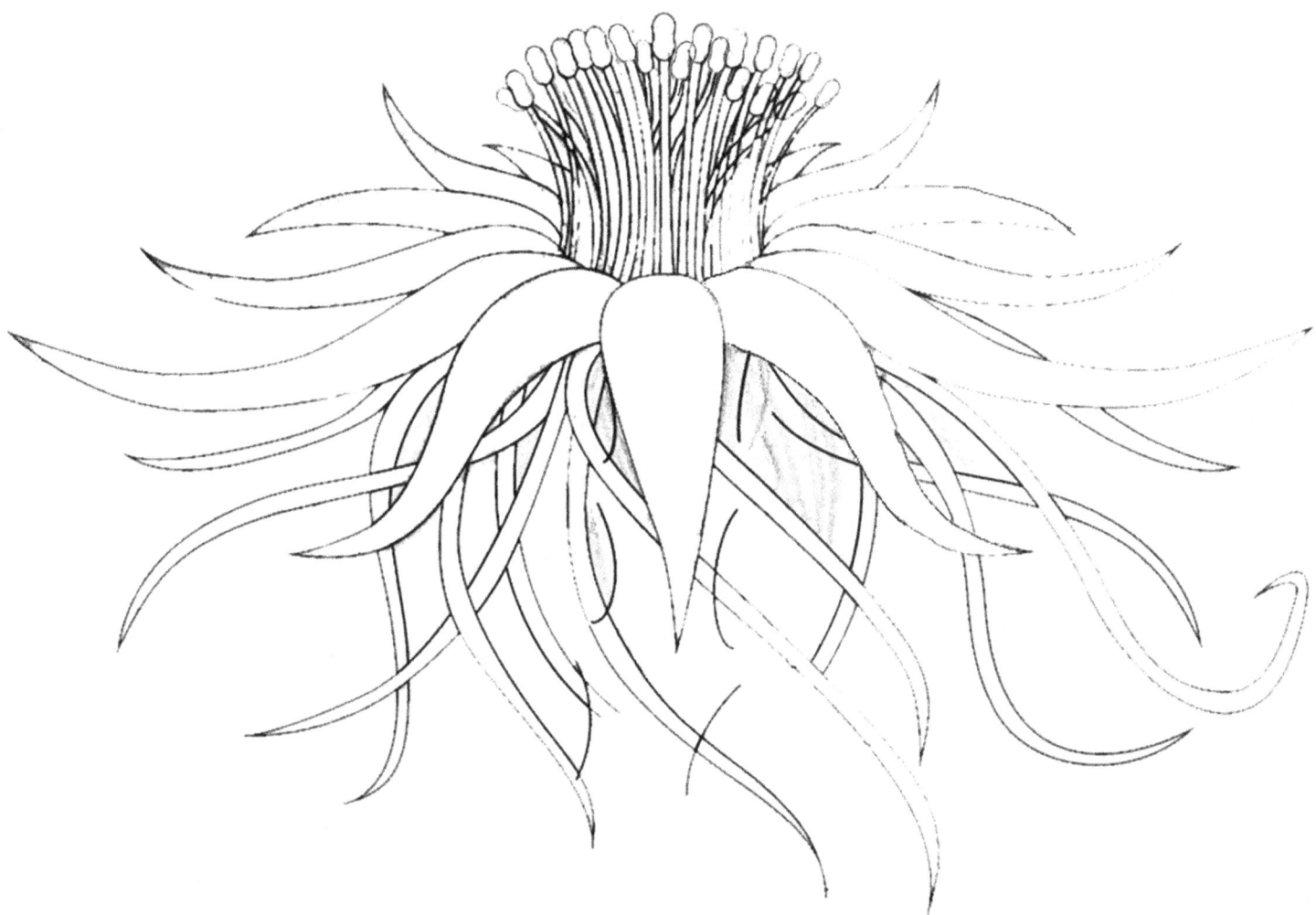

Color the Queen of the Night

Dedication

Tohono Chul is home to the nation's largest Cereus collection;
The garden grounds offer an immense selection.
Of the four hundred plants, half are native and natural,
Like the ones still found in the Tucson Mountain chaparral.

The Cereus scientific name is *Peniocereus greggii*
And honors an explorer, I'll tell you why.
The plant was described by Josiah Gregg, a naturalist
Along with some other plants not previously known to exist.

Acknowledgments

When Linda Wolfe of the Tohono Chul Botanical Gardens in Tucson suggested the night-blooming Cereus as a book topic, I was thrilled. That was until I tried to find details on these fascinating plants. Lee Mason, the Tohono Chul expert came to my rescue, providing an incredible amount of information. I am happy to share a tiny bit of his knowledge. Any mistakes are solely mine.

Many people helped me prepare the text: Lori Bonati, Susan M. Oyler, Pamela Bickell, Kate J. Steele, Brad Peterson and the Tucson Poetry Society. I want to thank Nick Thorpe for his enthusiasm in doing the illustrations. As always, I could never create the actual book without my editor, Nora Miller.

Elaine A. Powers, originally from Peoria, IL, currently resides in Tucson, AZ. After a career as a laboratory biologist, she is now pursuing her dream of writing science-based children's books and murder mysteries as well as continuing her work as a citizen scientist for iguana conservation. Her iguanas and tortoises continually inspire her.

For more information, visit her website www.elaineapowers as well as the publisher's webpage www.lyricpower.net.

More Books by Elaine

Bahamian Boa: a Tabby Tale
Tabby and Cleo: Unexpected Friends
Grow Home, Little Seeds!
Clarissa Catfish Visits the Peoria River Museum
Curtis Curly-tail and the Ship of Sneakers
Curtis Curly-tail Hears a Hutia
Curtis Curly-tail is Lizardnapped!
Fly Back to the Brac, Brian Brown Booby
How (Not) to Photograph a Hummingbird
Lime Lizards Boys and the Ship of Sneakers

Don't Call Me Turtle!
Don't Make Me Fly!
Don't Make Me Rattle!
Dragon of Nani Cave
Silent Rocks

And several iguana identification booklets and audio theatre scripts.

Nicholas Thorpe, Illustrator - Nick Thorpe lives in the desert where he does graphic design for some fanciful characters. Send water.

www.ingramcontent.com/pod-product-compliance
Lightning Source LLC
Chambersburg PA
CBHW051802200326
41597CB00025B/4658